MW01534176

BIONICS

Bionics

FRANKLIN WATTS
NEW YORK | LONDON | 1978
AN !MPACT BOOK

BY MELVIN BERGER

Photographs courtesy of: New York Public Library
Collection: pp. 7, 8; Wide World Photos: pp. 16,
33, 62; United Press International: p. 23; J.E. Wes-
cott, U.S. Atomic Energy Commission: p. 30;
Montefiore Hospital Medical Center: pp. 34, 41;
Bionic Instruments, Inc.: p. 48; Twentieth Century-
Fox Film Corporation: p. 69; Argonne National
Laboratory: p. 74.

Library of Congress Cataloging in Publication Data

Berger, Melvin.
    Bionics.

    (An Impact book)
    Includes index.
    SUMMARY: An introduction to the science that
is concerned with providing new parts for the human body, such as artificial limbs, organs, senses,
and even intelligence.
    1. Prosthesis — Juvenile literature. 2. Bionics —
Juvenile literature. [1. Prosthesis. 2. Bionics] I. Title.
RD130.B47   617'.307   77-17073
ISBN 0-531-01354-5

CONTENTS

Bionics Bionics
Bionics Bionics

INTRODUCTION INTRODUCTION INTRODUCTION INTRODUCTION

THE STORY OF BRENDA

Brenda, a high school senior, was out shopping one day on a busy downtown street. Just as she was passing in front of the large glass windows of a bank office, someone flung a paper-wrapped box from a passing car toward the bank. The bundle landed at her feet. Suddenly there was a powerful explosion. The impact hurled the girl a distance of 10 yards (9 m.) through the windows of the bank.

Brenda was rushed to the hospital and taken at once to the operating room where a team of surgeons attempted to save her life. After a few tense hours, they were satisfied that her brain and vital organs were functioning satisfactorily. But the right side of her body had been completely mutilated. The doctors repaired the organs and bones as best they could. But her right arm and leg, her right eye and ear, and many of the bones on that side of her body could not be saved. Their only hope was to use replacement parts.

Over the following three months, the surgeons performed six more operations. They replaced her smashed ribs and other broken bones with Vitallium metal, which was carved to the exact measurements as her natural bones, but was much stronger. For her amputated right arm and right leg, they attached artificial limbs powered by nuclear energy. To replace her right eye, they implanted a natural-looking glass eye. Inside the glass eye they hid a sub-miniature television camera. They fixed in place an ear that an artist had constructed and painted to look like her natural ear. In the ear opening, the surgeons placed a very sensitive, super-small microphone.

While she was still in the hospital, Brenda had many difficulties in getting accustomed to these new artificial parts. When she shook hands with people they winced with pain from the strength of her grip. Once when she was writing a letter, she pressed so hard that she pushed the pen right through the top of the desk. Often when she would try to push or kick open a closed door, she would put her arm or foot right through it. In tying scarves around her neck, she had to be careful not to strangle herself. And she had no trouble hearing whispered conversations held far from her room.

By the time she got home from the hospital, Brenda had learned how to use the added strengths and abilities of her new body parts. The day Brenda overslept and missed the school bus, she ran the four miles (6.4 km.) to school in exactly ten minutes. And, once, when her tiny pet parakeet flew out of its cage, and then out of the window, she was able to hear its call from a shopping plaza about eight streets away. She spied the bird in the sign on top of a building. With one leap, Brenda got up on the roof and rescued her pet.

But it was in school that Brenda most enjoyed her synthetic limbs and organs. In gym, of course, she could throw and kick farther, catch better, do more difficult gymnastic tricks, and win at almost any sport she tried. In biology she could see cells without a microscope. In her craft class she could bend metal, hammer nails, and chip stone by hand alone. And in all her classes she could read an entire page at a single glance.

THE PROMISE OF BIONICS

Brenda is part flesh and blood, and part metal, plastic, glass, and wire. She is also part fact and part fiction.

It is a fact that artificial limbs, organs, and bones already exist to replace those lost through disease or accident. But some of Brenda's replacement parts are still only ideas in the imagination of the scientists who are doing research to develop body parts that more closely imitate, or even surpass, the original parts. Both those that exist already, and

(4)

those that are yet to come, belong to that most exciting, most promising science, bionics.

The word bionics was coined by Dr. Jack E. Steele, of the U.S. Air Force, from the Greek word, *bion,* which means life or living. Dr. Steele first publicly defined bionics on September 13, 1960, at a scientific meeting called by the Air Force at the Wright Patterson Air Force Base in Dayton, Ohio. He described it as the study of the systems and structures of living animals and plants, and the application of these principles to devising machines and artificial systems for the benefit of humans.

The science of bionics is concerned with providing new parts for the human body, such as artificial limbs, organs, senses, and even intelligence. It is also involved with inventing and building machines and devices that can help people in other ways. Boats that move through the water as smoothly as fish, aircraft that fly as well as birds, and an ability to navigate as well as migrating animals are among the goals of bionics.

Bionics is a "bringing together" science. Biologists are part of the bionics' team. They study the animal and plant systems that provide the models for all developments in the field. Engineers contribute by conceiving and constructing the machines and hardware that realize these natural models in synthetic devices. In the slang of bionics, biologists do the wet work; the engineers do the dry work.

Other scientists have roles to play in bionics as well. Physicians or doctors are most important both in studying the human body and in preparing parts to be used in or on the body. Mathematicians translate the observations of the natural systems into the precise numbers and figures of a scientific model. Experts in electronics, nuclear energy, chemistry, the science of materials, and many other specialities contribute their expertise.

THE HISTORY OF BIONICS

According to an ancient Greek legend, Daedalus fashioned bird-like wings made of feathers and wax in order to fly and escape from prison with his son, Icarus. With these appendages attached to his arms, Icarus was indeed able to soar up into the air. Icarus' downfall, and the first failure of bionics, at least in fiction, was that he flew too close to the sun. The wax in his wings melted, and he crashed into the sea.

The oldest actual account of the use of a bionic device dates back nearly 2,600 years to the land of Persia around 600 B.C. A soldier, Hegesistratus, was imprisoned in the stocks. To escape, he cut off his foot and part of his leg. In a very simple application of bionics, Hegesistratus attached a wooden peg to the stump of his leg so that he could run away.

The most ancient bionic-like device still in existence was found in 1858 at an archeological dig in Capua, Italy. It is an artificial leg made of copper and wood that is believed to date back to around the year 300 B.C.

A Norse legend, thought to be from the fifth century A.D., tells of a flying machine made of feathers. An inventor named Wayland built this early aircraft, and his brother Egil was the first to try to ascend in it. Although the story tells us that he was able to fly, it fails to tell us how.

In the fifteenth century, the great genius Leonardo da Vinci devised flapping wings for flying. The design was based on his studies of the structure of a bird wing. Brilliant as it was, Leonardo never succeeded in flying in his invention.

The legendary flight of Daedalus and Icarus was the first failure of bionics.

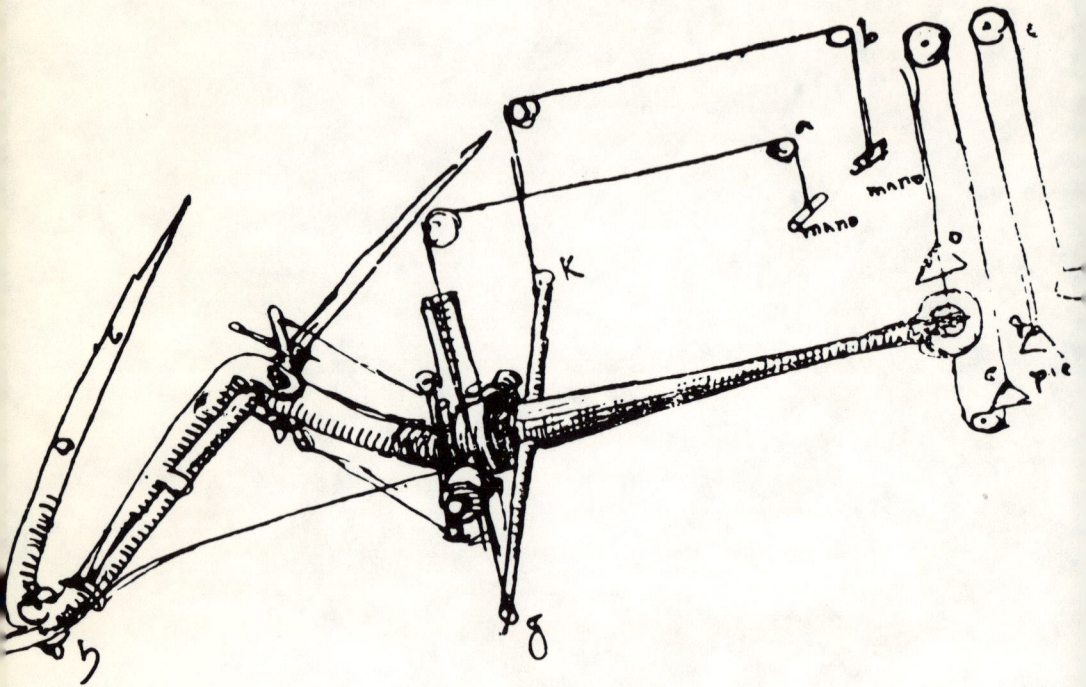

Leonardo da Vinci designed a bionic device to attempt flight. It was his greatest disappointment that it never succeeded.

The pace of advances in bionics picked up in the eighteenth century. In colonial America, Paul Revere was famous for his midnight ride, his silversmithing, and the false teeth he made of animal bones. Equally famous are the wooden false teeth of George Washington. And Benjamin Franklin is credited with the invention of eyeglasses.

In 1790, while building the first steam engine, James Watt attached a device to prevent the machine from spinning too fast. He called it a governor, from the Greek *kybernetike* and the Latin *gubernator*. Both refer to the helmsman who steers a ship. Just as the helmsman adjusts the wheel so the ship stays on course, so the governor controls the steam engine so it spins at the desired speed.

A governor depends on a process called feedback. Feedback compares the results of an action with fixed reference points. For example, the ship's engine can spin as fast as 100 times a minute, but any speed over 80 times a minute is dangerous, or the ship may head due north when the course should be north by east. Using feedback, if the engine is spinning too fast, or the ship is off course, the mistake is corrected, either automatically in the steam engine, or by hand on the ship.

During the nineteenth and early twentieth centuries, there were steady improvements in the human applications of bionics—better artificial limbs, organs, eyeglasses, hearing aids, dental devices, bones, and other replacement body parts.

There were also widespread applications of bionics in other areas. The discovery that bats use high-pitched cries and echoes to guide their flight led scientists to develop radar and sonar. The adaptive camouflages that help many animals to avoid detection by blending in with their surroundings inspired the military to devise clothing and coverings that soldiers could use to conceal themselves and their equipment. From the squid that squirts ink into the water to protect itself against a predator came the idea of using smoke screens to hide and protect troop movements.

Long-term research projects of every kind showed scientists, among other things, how birds and fish are able to navigate for long distances, how birds can fly with their own muscle power, how snakes smell and detect heat, how certain plants live without water, how frogs can see and catch insects, as well as the extraordinary survival abilities of other animals.

CYBERNETICS

The science of cybernetics, launched in 1947 by Professor Norbert Wiener of the Massachusetts Institute of Technology, is an important forerunner of bionics. According to Professor Wiener, cybernetics is the science of control and communication. In these studies scientists try to understand human and animal actions in mathematical and logical terms.

The first goal of cybernetics is to construct a theory that can explain elements of human actions. This theory is based on finding the answers to questions such as: How do humans learn? How do they use past experiences to cope with new situations? What steps do people follow in solving problems? The greatest progress in cybernetics is in this area.

The second goal is to construct a model that shows the steps that people follow in arriving at a conclusion. Already cybernetic research has helped to express particular human thoughts and problem-solving abilities in mathematical or machine-like terms.

The third and most difficult goal of cybernetics is to construct a system that follows a step-by-step approach to problem-solving. This type of system usually involves a computer. Since cybernetics is trying to understand highly complex aspects of human actions, this goal is the hardest to accomplish. Meanwhile, cybernetics is making progress in understanding human thought and is bridging the gulf between people and machines.

RECENT HISTORY OF BIONICS

In 1951, the U.S. Navy, through its Office of Naval Research, undertook a research project to study living organisms, animal and plant, to see if they suggested models the Navy could use in creating new mechanical or electrical systems for their ships or submarines.

Some years later, the Air Force started a similar project. At the 1960 meeting at which Dr. Steele used the term bionics for the first time, 100 biologists, engineers, mathematicians, physicians, and psychologists were in attendance. These scientists were involved in bionic research, supported by grants totaling $3 million.

In only three years, at the second bionics meeting, 1,000 scientists representing the various disciplines had become involved. Their research was funded by new grant monies that amounted to about $15 million.

The 1970s have seen a change of direction for the old-new science of bionics. Scientists in the field who had been devising systems that imitate complicated patterns of animal thought and actions are returning to the parent science, cybernetics. Building the hardware is less important to them than getting a basic understanding of thinking and doing.

The bionic scientists who are interested in creating machines and things to help people handicapped by missing limbs or poorly functioning organs, are now part of the broader science of biomedical engineering. Besides bionics, this science applies various advanced technologies to medicine: the use of X rays and nuclear energy to diagnose and treat disease; devising modern electronic equipment to measure and monitor the heart, blood, and breathing, as well as the electrical and chemical changes that occur in the body.

(11)

It brings together biology and medicine on the one hand, and mechanical, electronic, and chemical engineering on the other.

In the bionics branch of biomedical engineering, some researchers are interested in the successful replacement of missing limbs. They want artificial limbs to look and function as much like natural limbs as possible, if not better. Others are studying the replacement of entire organs, or parts of organs, that have been damaged by disease or accident. Still others are interested in improving or restoring one or more of the senses, especially those of sight and hearing. And, finally, there are those working in artificial intelligence who want to find ways to extend and improve people's ability to think.

Few people today call themselves bionicists, and few colleges offer courses in bionics. Yet this specialized science is stretching people's imagination and challenging their ingenuity as they probe and try to imitate the wonders of nature.

ARTIFICIAL LIMBS

ARTIFICIAL LIMBS

ARTIFICIAL LIMBS

ARTIFICIAL LIMBS

THE STORY OF JAY ARMES

Even as a young boy, Jay Armes was very ambitious. The son of a wealthy grocer in El Paso, Texas, Jay held three part-time jobs and ran an active loan service during lunch time in school. By the age of 12, he had already earned a sizable amount of money.

Then Jay had a terrible accident. Two sticks of dynamite exploded as he held them in his hands. Nearly all of both hands were blown away in a shower of blood and flesh. The surgeons had to amputate at about 2½ inches (6.3 cm.) above the wrists.

Jay was fitted with crude hooks to replace his missing hands. Less than a month after the accident, Jay was back in school. He quickly resumed a full schedule of in-school and after-school activities. Even with his artificial hands, Jay was chosen for the school football, baseball, basketball, and boxing teams. He graduated at the age of 15.

At college, Jay studied criminology and psychology. After earning degrees in these fields, he returned to El Paso. He rented a small office and opened a private detective agency. He specialized in tracing missing persons—either victims of kidnapping or those who had run away.

Over the years Jay managed to solve every case that he accepted. As his reputation grew, so did his fees. He became a very successful and wealthy man.

Jay Armes still works in the same small office in El Paso. But he now lives in a mansion worth $1 million. The house is equipped with a gymnasium where Jay works out daily and practices karate, a shooting range where he practices marksmanship, and a space for him to keep a menagerie of wild animals.

Armes' investigative activities have brought him many enemies. He has had over a dozen serious threats on his life. He protects himself with an electrified fence around his

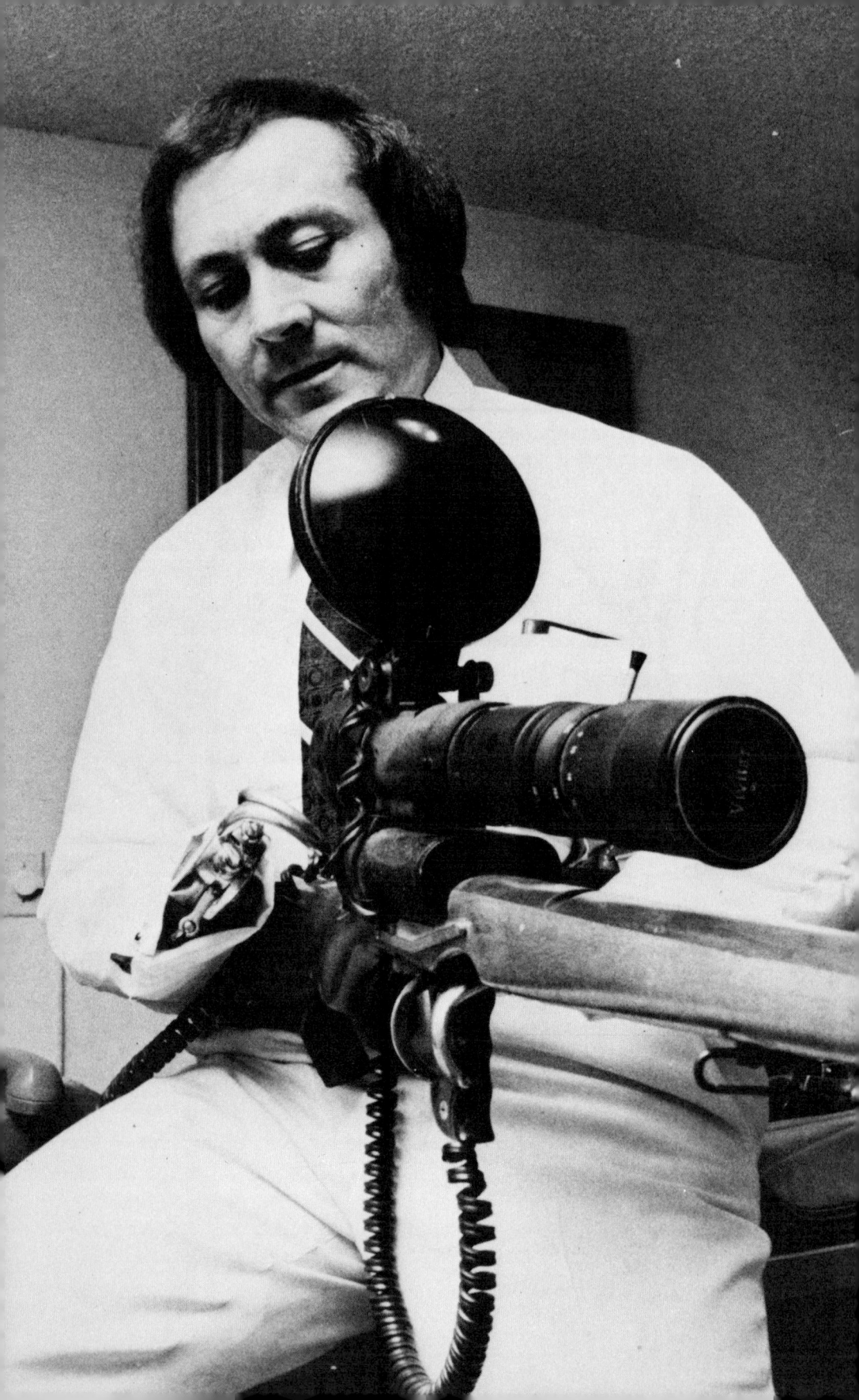

house and property, armed bodyguards, and the wild animals that are allowed to roam the grounds freely at night.

By now Armes has the best artificial hands that money can buy. In addition, he had a $2,000 Magnum pistol built into his right hook. To fire the gun, he merely flexes the muscles in his right arm. The gun has helped him to survive more than one attack. Small wonder that so many are curious about the $10,000 device that he had built into his left hook, and that he refuses to describe.

ARTIFICIAL ARMS AND HANDS

The science of making an artificial body part to replace a missing or damaged one is called prosthetics. The actual artificial part is called a prosthesis (plural, prostheses). Bionics helps to make arm and hand prostheses that closely resemble the function and appearance of natural limbs. They even attempt prostheses that in some limited ways are better than natural limbs.

All artificial replacements of the arm should have a few essential capabilities. Most important is that it be able to bend at the elbow. The bend of the wrist is considered less important.

As for the hand, it is most important that it be able to grasp objects, including objects of irregular shapes and of different sizes. The natural hand grasps by using the thumb in apposition to the other fingers. The soft pads on the finger tips conform to the shape of the object, and maintain the grip by friction.

Jay Armes overcame his physical disability with the use of artificial hands.

In its simplest form, an artificial limb is nothing more than a wooden or plastic device shaped to resemble the natural arm. It has a hinge arrangement that works somewhat like the natural elbow. This works by having a strap go from the artificial elbow joint to the harness that holds the arm in place. The harness passes tightly across the wearer's shoulder and back. By tightening and relaxing his or her shoulder muscles, the wearer moves the strap, bending the prosthesis at the elbow.

The most basic functional hand prosthesis has a metal hook. If you look closely at a prosthetic hook, you can see that it is actually made of two separate hooks pressing against each other. They are both made of heavy steel, shaped like big question marks.

One of the hooks is permanently fixed in place. The other one is attached to a spring mechanism that keeps it pressing against the unmoving hook. A cable runs from this hook to the prosthetic harness. The wearer uses his or her back muscles to raise the movable hook and separate the two hooks.

The wearer can grasp small objects between the hooks. They can be used for eating, dressing, writing, driving, telephoning, and similar activities. Or they can be used as a unit, just like the natural hand, for carrying and lifting.

A surgical procedure first used during World War I, called cineplasty or kineplasty, allows for even more efficient control and use of the artificial hand. In this operation the surgeon implants steel pins in the muscles remaining in the stump of the arm. Cables run from these pins to the artificial hand and arm.

To move the hand, the user has merely to think about the movement he or she wants to make. A nerve impulse travels from the brain through the nerves to the muscles that control that movement. The muscles then automatically shorten or lengthen. This either tightens or loosens the cables that are

(18)

attached to that particular muscle. The changes in the cable length move the artificial arm and hand, resulting in movement similar to that of the natural arm and hand.

Most recently, some artificial limbs have been built which can be moved directly by nerve impulses from the brain. Tiny electrodes, placed either on the skin of the stump or in the muscles themselves, pick up the signals from the brain to the muscles. The signals pass through wires to a miniature electrical motor. This motor is powered by batteries that are usually carried in a pack held on the wearer's belt. The motor actually moves the joints and fingers of the prosthesis.

Devices that are powered by the electricity in the muscles are said to be myoelectric. The word comes from the Greek, *myos,* which means muscles. Future developments in artificial limbs will most likely be based on new uses of myoelectricity.

In 1975, Reid Hilton, a 24-year-old karate expert from Santa Ana, California, was one of the first people to receive an advanced, myoelectric arm after he lost his right arm below the elbow in an accident. This experimental arm weighs only 8 pounds (3.6 kg.). It is connected, not to the muscles, but directly to the nerves that bring the impulses to the arm.

The doctors permanently inserted three small buttons through the skin in his arm. One button was connected to the arm's median nerve, another to the ulna nerve, and the third was set up as a ground to prevent electrical shocks. Wires go from each of the three buttons to a power pack inside the hollow prosthetic arm. The power pack includes an amplifier and a motor to control the movements of the hand and fingers.

With his myoelectric arm, Hilton thinks of the movements he wants to make and they are made. He can pick up tiny objects, tie his shoe laces, screw in a light bulb, fold a letter and insert it into an envelope, and open a pack of cigarettes. While the average male has a hand grip of about 25 pounds (11 kg.), Hilton's artificial grip is measured at 40 pounds (18 kg.). And he can maintain the pressure for hours without tiring!

This artificial arm is still in the testing stage. It costs about $40,000 to make, while one of today's ordinary artificial arms costs between $1,500 and $2,000.

Scientists are working to develop improved prostheses. They are searching for new materials that will be lighter and stronger than the ones presently in use. They are testing better designs that will more closely resemble natural limbs. They are trying to build motors that will be even more sensitive to myoelectric signals, and yet stable enough to resist the burst of electricity produced by a sneeze! They hope to find power sources more efficient than batteries, which tend to be heavy, awkward to carry, and short-lived.

One promising device weighs 2 pounds (0.9 kg.) and uses fluids and hydraulics to control movement. The device moves the arms and fingers by using a system modeled on a natural limb.

The motor is a small, high-speed electrical pump which is controlled by myoelectric signals. When the brain signals that it wants to bend the arm or a finger, the motor is actuated. A spurt of fluid is forced into several thin, flexible tubes. The fluid causes the tubes to expand and thereby become shorter, the way that natural muscles contract when they receive a signal from the brain.

Researchers are also working to build artificial limbs that will be sensitive to pressure and even to temperature. Right now, the wearers of ordinary artificial limbs complain that the prosthesis is like an arm or leg that has fallen asleep. They cannot tell how hard they are pressing or the temperature of the object they are holding.

The scientists hope to give feeling to the artificial limbs

by using sensors. Sensors are devices that detect energy in one form and change it to another. A thermometer that changes heat to movement, and a phonograph that changes movement to sound are examples of sensors.

The sensor used to detect pressure is usually the strain gauge. When anything presses on the strain gauge, it produces a small current of electricity. The greater the pressure, the greater the flow of electricity. Part of this electricity is fed back to the motor in a feedback arrangement. The motor automatically adjusts to any changes in electrical flow. If the flow of electricity goes above a certain limit, the motor immediately slows down, releasing the pressure. If the flow of electricity is below the set level, the motor speeds up and increases the pressure.

Scientists are also trying to develop a device that will be as useful as the hook, and at the same time as attractive as the cosmetic hand. One possible solution is an artificial hand with three movable parts, corresponding to the thumb, index finger, and middle finger. These three metal fingers are set in a material similar to foam rubber that is sculpted to look like the user's natural hand. The entire hand is covered with cloth, and has a layer of tiny glass beads under the cloth. In some models there are inflated pads on the finger tips. Early tests show that this hand is able to grasp and hold almost as many objects of different size and shape as the natural hand.

Another important direction is toward the use of nuclear energy—instead of batteries—as the power source for artificial limbs. A small amount of a radioactive substance, such as plutonium, is placed inside the prosthesis. It must also be kept within a sealed metal container to protect the wearer and others from the dangers of exposure to radiation.

As the element goes through the radioactive decay, it gives off heat. A thermocouple, made of two different metal wires joined together, produces an electric current when the joint is heated. The electricity from the thermocouple may then be used to power the electric motor that controls the hand and arm movements.

THE STORY OF TEDDY KENNEDY, JR.

Young Teddy Kennedy, nephew of the late president, had a rare form of cancer in the bones of his right leg that was discovered when he was 12 years old. To prevent the spread of the disease the surgeons had to amputate his right leg, four inches (10 cm.) above the knee.

Directly after the one-hour surgery, a surgically scrubbed, gowned and masked biomedical engineer stepped up to the operating table. He was a specialist in designing and fitting artificial limbs. For the next 20 minutes he worked to apply a tight, stiff plaster cast over Teddy's thigh, and over the wound. To the lower part of the cast he attached a temporary artificial leg. It was made of metal tubing, with a foot of rubber.

When Teddy awoke several hours later, the shock of loss was somewhat tempered by the artificial leg that was already in place. The next day the boy got out of bed, and was able to stand up. In three days he was walking around on his artificial leg. Each day he worked at walking and performed exercises to build up the muscles needed to move the leg. Within two weeks he could get around very well.

One month later the same biomedical engineer fitted Teddy with a permanent limb. It was the same size, shape,

Teddy Kennedy, Jr. is a master at skiing on one leg, without his prothesis.

(22)

and shade as his natural leg. It bent at the knee and ankle. At first Teddy could only walk with crutches. After a short period of adjustment, however, he was able to get around without using crutches or a cane. And, in time, not only was Teddy able to walk on the leg, he was also able to play football, baseball, sail, ride his bicycle, and participate in all the other sports he had always enjoyed. Teddy even learned how to ski without the artificial leg at all by using special poles with short skis at the ends instead of points.

Although his battle with cancer is still not over, the artificial limb allows Teddy to lead a normal, active life.

ARTIFICIAL LEGS

The procedure used on Teddy Kennedy, Jr., was first introduced in 1963. Thus far it is the most modern method of fitting an artificial leg. It is a big improvement over earlier kinds of prosthetic limbs that had been used. This type of limb is put on earlier after the amputation, fits more snugly, and does not require the wearer to use heavy belts or other clumsy kinds of attachments.

In the past the doctors allowed the wound to heal after the amputation before fitting the prosthetic device. They believed the post-operative swelling had to go down before the artificial limb could be properly fitted. Today a heavy cast is placed over the stump immediately after the operation, and a temporary limb is attached at once. Two weeks later this cast is taken off, which allows the surgeon to remove the stitches. Another, lighter cast is then put in place for two weeks.

When a final, light cast is put in place, a cast of the stump is made. It is sent to a factory to start the manufacture of a carefully fitted, permanent limb. In a little while the patient is fitted with the new limb. Most wearers, using crutches, are able to walk out of the hospital on their new legs. Though it is difficult to walk at first, most patients are so strongly motivated that they learn in a week or two.

The artificial leg is held in place by suction. Therefore the wearer who participates in active sports may need to use a light belt to make sure that the vacuum is not broken.

From the viewpoint of bionics, the main parts of the natural leg are the strong bones and muscles, the joints at the knee and ankle, and the flat surfaces of the soles of the feet. The bones support the body, and the muscles are necessary for the person to be able to move his or her legs and feet. The flexibility of the knee and ankle joints allows the person to go through complicated walking and running movements. And the flat bottom surface supplies friction to prevent slipping, and supplies stability to prevent toppling over.

Prosthetic manufacturers already can supply artificial legs with any one of 150 different types of knee joints, and 50 different types of ankle joints. Some are simple hinge and lever devices. Others work on hydraulics or compressed air. All try to duplicate the movements of the natural legs and joints as closely as possible. For youngsters, new, bigger limbs are fit from time to time to keep up with their growth.

Now that progress is being made in the mechanical development of the artificial leg, users are asking that it be made more natural-looking. The latest design covers the plastic leg with a thin layer of flesh-tinted foam rubber. The developers are still working out the problem of protecting the foam rubber from tearing, and of maintaining its shade and shape.

Great advances have been made in designing, building, and fitting artificial limbs. But new strides in technology are expected to make even greater improvements possible in the near future.

ARTIFICIAL ORGANS

ARTIFICIAL ORGANS

ARTIFICIAL ORGANS

ARTIFICIAL ORGANS

ARTIFICIAL ORGANS

ARTIFICIAL ORGANS

THE STORY OF ANTHONY J. TASCO

At 35 years of age Anthony J. Tasco developed a condition that interfered with his ability to work at his job as a building worker. He felt very tired or nauseous much of the time. He experienced spells of dizziness and fainting. Several times he blacked out entirely.

Tony's disease was diagnosed as heart block. The muscles of his heart, it was found, were not pumping blood regularly through his body.

The doctors tried to treat Tony's condition with drugs. He was in and out of the hospital many times. But there was no improvement; his heartbeat continued to be weak and irregular. For several years Tony stayed home, unable to work, leading the life of an invalid.

Finally it was decided that Tony should have an operation. The surgeons implanted a nuclear-powered pacemaker into his chest. A pacemaker is an electronic device that establishes a normal, regular pattern of heartbeats.

Tony was in the hospital for about a week while he recovered from the operation. Then the doctors told him to take it easy for a while at home. Finally they told Tony he could go back to work and resume almost all of his normal activities—with the nuclear pacemaker controlling the beating of his heart.

HELPERS AND SPARE PARTS
FOR THE HEART

Heart disease is the major killer and crippler in America and Europe today. More than half the deaths in the United States are caused by heart disease. But new prosthetic devices and new surgical procedures are now correcting many cases of heart disease and preventing death from heart failure. Tony is one of many hundreds of thousands who have benefited from these life-saving advances.

(29)

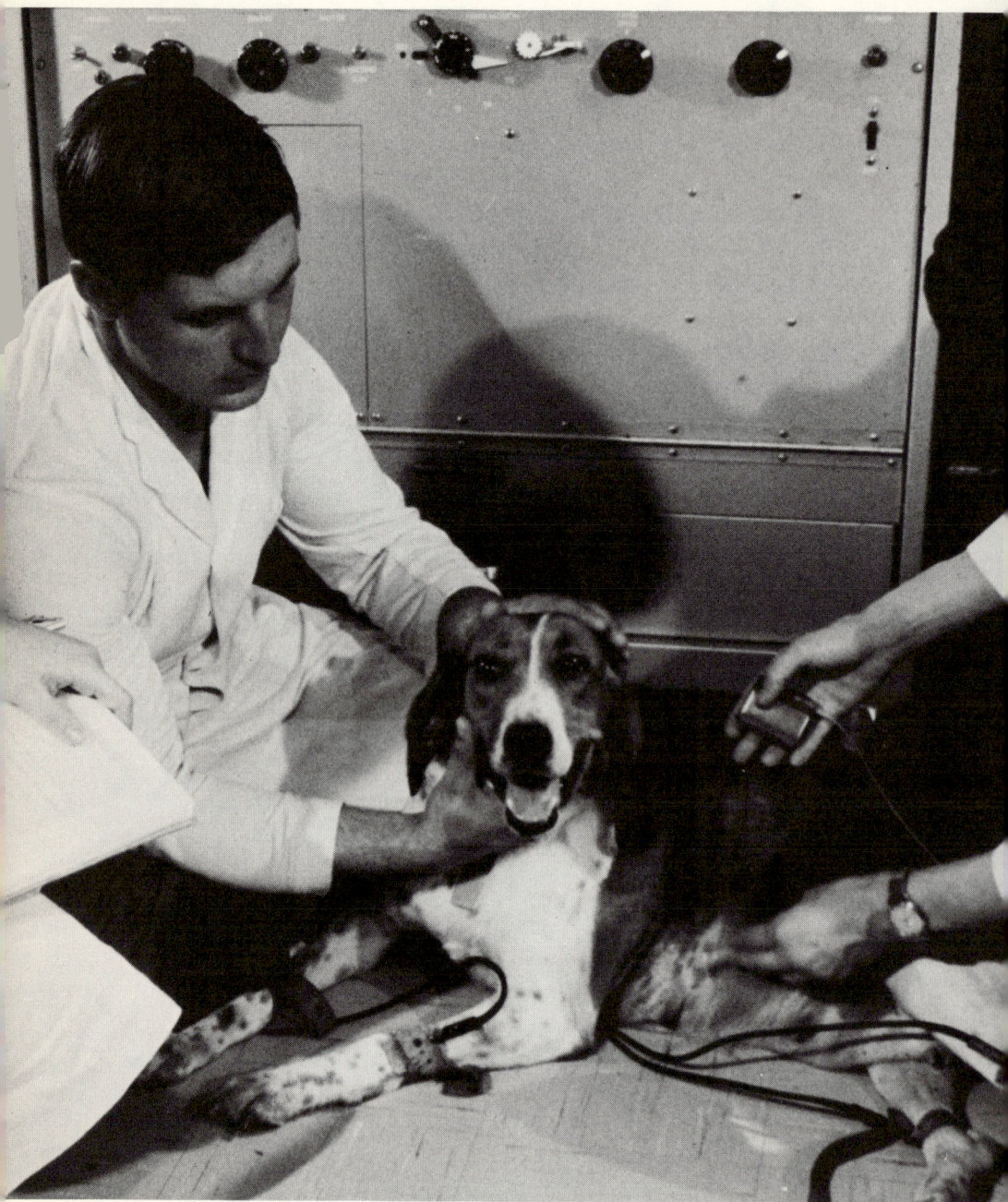

Essentially, the heart is a big muscle. It is a powerful pumping engine that sends the blood coursing through the arteries of the body. As the blood travels, it brings the necessary oxygen to all the tissues and organs, and picks up the waste carbon dioxide that they are releasing.

The blood then returns through the veins to the heart. The heart pumps the blood, now laden with carbon dioxide, to the lungs, where it exchanges the carbon dioxide for oxygen. And then, again, the heart sends the blood out through the arteries of the body.

As long as your heart pumps blood, your body gets the oxygen that it needs. Normally, the heart beats automatically and regularly. Sometimes, however, the rhythm becomes irregular, due to faulty functioning of the nerves of the heart that regulate rhythm. The heartbeat slows down or speeds up, becomes very weak or even stops for a few seconds. This interrupts the supply of oxygen to the organs, and particularly to the brain. A few minutes without oxygen can damage your brain; longer can cause death.

Heart block is a disturbance in the heart's conducting system. The chambers of the heart beat at different rhythms. In cases of heart block that are not helped by drugs, the treatment is to provide electrical pulses to the heart from an outside source, such as an artificial pacemaker.

Starting in 1952, doctors began experimenting with sending the electrical impulses into the heart from outside the body. In time the source of electricity was made small enough for it to actually be implanted inside the body in a minor surgical operation. The original models were powered by

**Brunhilde the beagle has
been fitted with a
nuclear-powered pacemaker.**

(31)

batteries, also in the body, which ran down in two years and had to be replaced. This required another operation.

Tony Tasco's operation, in 1972, was important because he was the first American to have a nuclear powered pacemaker implanted in his chest.

Tony's pacemaker is about as large as a cigarette case, and only weighs 4 ounces (113 gm). Four hundred milligrams of plutonium 238 emit radioactivity continuously. The radioactivity produces heat. The heat is converted into electricity. It is expected that this amount of plutonium can power the pacemaker for at least ten years.

Plutonium rays, of course, are extremely dangerous. But in the pacemaker the plutonium is sealed in a titanium case. Therefore, the plutonium only gives off into the air as much radioactivity as the dial of a luminous watch. But what happens in case the person with a nuclear powered pacemaker is burned in a fire, drowned, involved in an air crash or car accident, or struck by a bullet?

During the testing of the titanium case, it was burned, immersed, smashed, banged, and subjected to all sorts of other abuse. The case held up to all these attacks. In fact, the only thing that broke the case was a rifle bullet fired directly at it from very close range. The chances of a person being hit with such a shot are very slim. Statistically, if 100,000 pacemakers are in use, the probability is that one of them will be hit every 42,000 years!

The pacemaker is usually implanted just below the skin of the abdomen or chest. The surgeons thread two wires from the pacemaker to the heart, and attach the wires to the walls of the heart. Seventy-two times a minute, tiny pulses

**Pacemakers have been success-
ful even in newborn babies.**

(32)

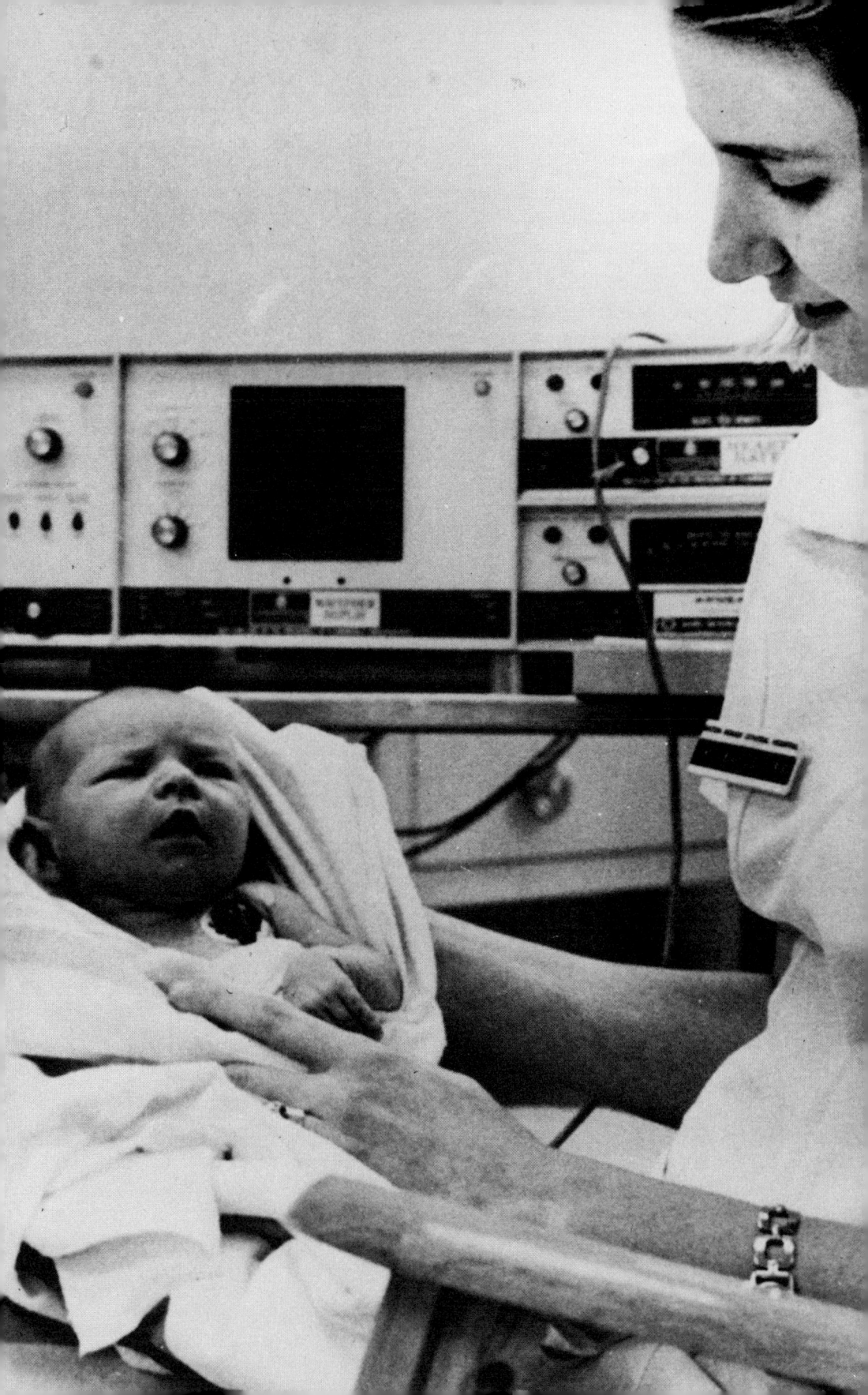

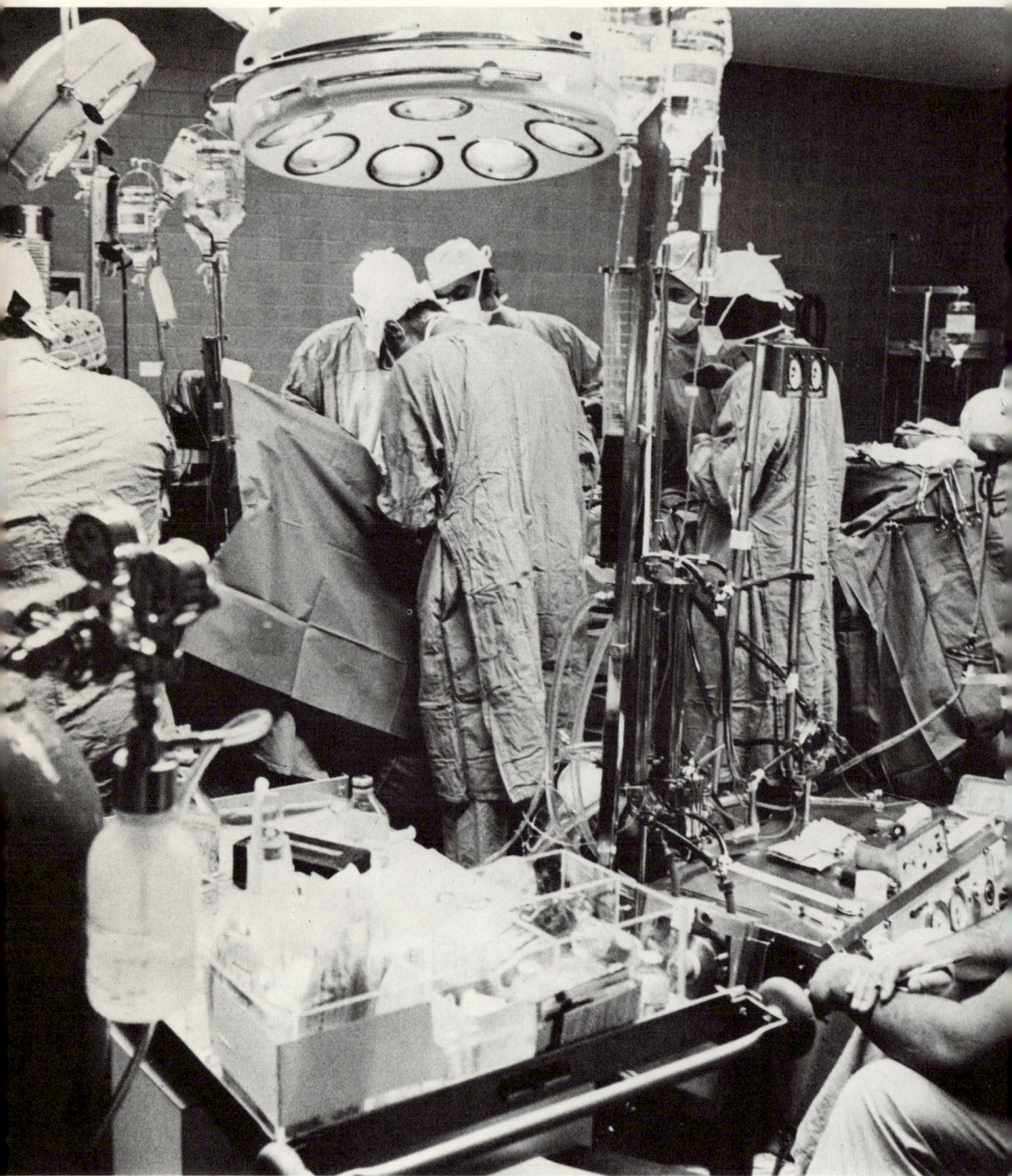

of electricity flow from the pacemaker, through the wires, and into the heart muscles, regulating the heart's rhythm.

Advanced models of the pacemaker have sensors on the wires that can determine whether the heart is able to beat regularly on its own. If it is, the pacemaker does not send out its pulses. But it is ready to send out its life-saving electricity as soon as the heart shows any signs of faltering.

Heart block cannot be helped by surgery. Surgery can, though, cure or improve some types of heart disease. For a long time, however, doctors could not operate on the heart. As soon as they cut into the heart tissue to repair or replace some damaged part, the heart stopped beating. And since it is fatal for the heart to stop for even a few minutes, the doctors were prevented from doing any open-heart surgery.

The invention of the heart-lung machine in the early 1950s made heart operations possible for the first time. This big, bulky machine is kept in the operating room. It is able to take over the work of the heart during surgery. And it can continue working for the length of the operation.

The doctor uses a thin plastic tube to shunt the patient's blood from the heart to the heart-lung machine. In the machine, a pump (similar to the heart) sends the blood through various tubes where the carbon dioxide is removed and the oxygen is introduced (similar to the lungs). Another pump then sends the blood, now rich in oxygen, back through a return tube to the patient's heart.

When the open-heart surgery is over, the surgeons disconnect the tubes from the patient to the heart-lung machine. They immediately give the heart an electrical shock to start it beating on its own.

**A heart-lung machine at work
during open-heart surgery.**

(35)

A frequent use of open-heart surgery is to correct faulty valves between the chambers of the heart. Repairing a damaged valve or replacing it with an artificial one gives victims of this type of heart disease a new lease on life.

Valves control the flow of blood through the heart. They are like one-way doors that let the blood flow from one part of the heart to another, but prevent it from flowing backwards. Faulty valves either do not allow enough blood through, or else they allow blood to leak back.

Defective valves that stick together, narrowing the valve opening, may be repaired or corrected in an operation. But leaking valves are usually replaced with artificial valves.

In the operation to replace a heart valve, the patient is connected to a heart-lung machine. The heart stops as the surgeon opens the heart, exposing the defective valve. The surgeon removes the valve and sets the artificial valve into place. It consists of a metal ring with a cage above it. Inside the cage is a ball or dish that can either fit snugly in the ring or float about in the cage. The doctors attach the new valve to the heart by means of the metal ring.

Now, when the heart beats, blood flows through the valve. The ball or disc floats up into the cage so the blood can pass without interference. But between beats, the ball or disc falls back snugly into place on the ring, preventing the blood from leaking backwards. In this way, the artificial valve lets the blood flow in one direction only.

Once a heart has become very weak as a result of disease, it cannot pump blood properly. One way of helping a failing or weak heart is to use a mechanical pump. Mechanical pumps were first successfully used by Drs. Michael E. De Bakey and Adrian Kantrowitz in 1965. These American surgeons invented the artificial ventricle. About the size and shape of an orange, this device has a tube that takes blood from the heart, and then, by air pressure, pumps the blood back through another tube into a large artery, relieving the strain on the heart.

More recently, scientists have begun working on models of a completely artificial heart. Dr. Willem Kolff, an outstanding scientist in the field of prosthetics, is a leader in this line of research. Thus far, all of his efforts have been in building animal hearts, particularly for calves. In 1973 he implanted the first artificial heart in a calf. The animal lived for one month. He implanted an improved model in 1975, and this calf lived for over three months. The third heart was implanted in 1976, and the animal survived for nearly five months.

Experiments are underway in Dr. Kolff's laboratories to develop a nuclear powered artificial heart that will keep an animal alive for one year. Only when he succeeds in these experiments will Dr. Kolff think of human experimentation.

For people who ask when an artificial heart for humans will be ready, Dr. Kolff always has the same reply: "I will be disappointed if the artificial heart is not ready in three years, and three years ago I said the same thing."

THE STORY OF DELBERT SEARLES

Delbert Searles was born prematurely. He weighed just over 2 pounds (1 kg.) at birth. For a few crucial moments after he was born, Delbert stopped breathing. This short period without oxygen permanently damaged his brain and caused spastic cerebral palsy. This is a crippling disease in which the brain is unable to control and direct the muscles properly. Delbert's left arm was permanently twisted, and his two legs were scissored across each other.

As the boy grew up and began to speak, he had difficulty in speaking clearly. Then, at the age of 5, he developed epilepsy, a disease of the brain. From time to time his muscles would suddenly convulse, and he would lose consciousness.

Between the crippling effects of cerebral palsy and the epileptic seizures, Delbert's future was very uncertain. When his mother learned of a new, but very expensive, operation to implant a brain pacemaker, she decided to run the risks of surgery in the hope that it might normalize her son's life.

A brain pacemaker directs an electrical signal to the brain, in much the same way as a heart pacemaker directs electrical signals to the heart. The brain pacemaker blocks the electrical signals that interfere with the ability of the brain to control the body's muscles.

Within a week after the operation, Delbert's arms and legs began to straighten out. With physical therapy treatments he was able to build up and strengthen unused muscles. Gradually he began to walk with help. Doctors expect that the little pacemaker will enable him, one day, to run and play like other children.

AID FOR THE BRAIN

The brain is the most complex organ in the body. It has many parts. Each part helps to control and direct different activities in your body.

The cerebellum is the part of the brain that automatically regulates the way you stand and move. Nerve pathways connect the cerebellum with the muscles controlling muscle tone, strength, and coordination of muscular movement. Damage to the cerebellum results in jerky, sometimes reeling and swaying movements, and loss of muscle tone necessary for maintaining posture.

To implant a brain pacemaker, surgeons cut two small holes, about 1 inch by 3 inches (25 by 76 mm.), in the back of the patient's skull, just above the neck. They attach electrodes to the cerebellum, and thread the wires under the patient's skin from the cerebellum to the front of the chest. Here they attach the wires to a tiny radio receiver, approximately the size of a large coin.

The patient wears a miniature radio transmitter and power source—such as a pack of batteries—around the waist. From the transmitter a wire leads to a very small radio antenna that is taped in place just over the receiver that is under the skin.

The radio signals travel from the transmitter to the antenna. They are picked up by the receiver and carried to the brain, where they are changed into electrical impulses. The user feels absolutely nothing as this goes on. These signals correct the poor pattern of signals that the brain is transmitting to the muscles. In some conditions the transmitter operates continuously, sending an average of 10 volts to the brain. In other cases, the patient turns the transmitter on when he or she feels an attack coming on.

Locating the power source outside the body makes it easy to replace batteries without surgery. Also, there is no break in the skin that might become infected. The brain pacemaker is easy to operate and to control, and offers patients relief from the crippling effects of certain kinds of brain damage.

A New York City father of three, Don Chiriani, was paralyzed from the waist down as a result of a serious spine disorder. He was also in constant pain, which no drugs were able to control. Over the years he had four operations on his back, but his condition worsened, and the pain increased. Finally his doctor suggested an operation to implant a painkiller similar to the brain pacemaker.

This tiny, battery-powered device is taped to the patient's back. It transmits pulses to an internal receiver with two electrodes placed on either side of the spinal cord. The electricity flowing through the electrodes blocks the pain signals set up by the damaged part of the back. In effect, the electricity from the transmitter neutralizes the electrical pain messages passing through the spine, thus eliminating the sensation of pain.

Most patients who receive these devices are suffering with pain so severe that they had been depending on huge amounts of painkilling drugs. The pain was so intense that they were usually confined to bed, and in danger of starving to death because they could not even eat. While these devices do not cure or correct the problem, they do help the patient to get out of bed and lead a more normal life.

AID FOR THE KIDNEYS

The kidneys are a pair of bean-shaped organs that have the important job of removing the waste products that the blood collects as it travels through the body. If the kidneys stop working, the poisons accumulate in the blood. Within a day or two, the person is sick. A few more days, and the person could die.

During World War II, Dr. Willem Kolff, mentioned earlier for his work in developing an artificial heart, built the first artificial kidney machine in his native Holland. It was huge, the size of an automatic washing machine. Although it was primitive compared to today's models, it set the basic principle on which all artificial kidneys work.

A kidney patient is plugged into the kidney machine two or three times a week, for four to eight hours each session. A

The kidney machine is saving thousands of lives that would once have been lost.

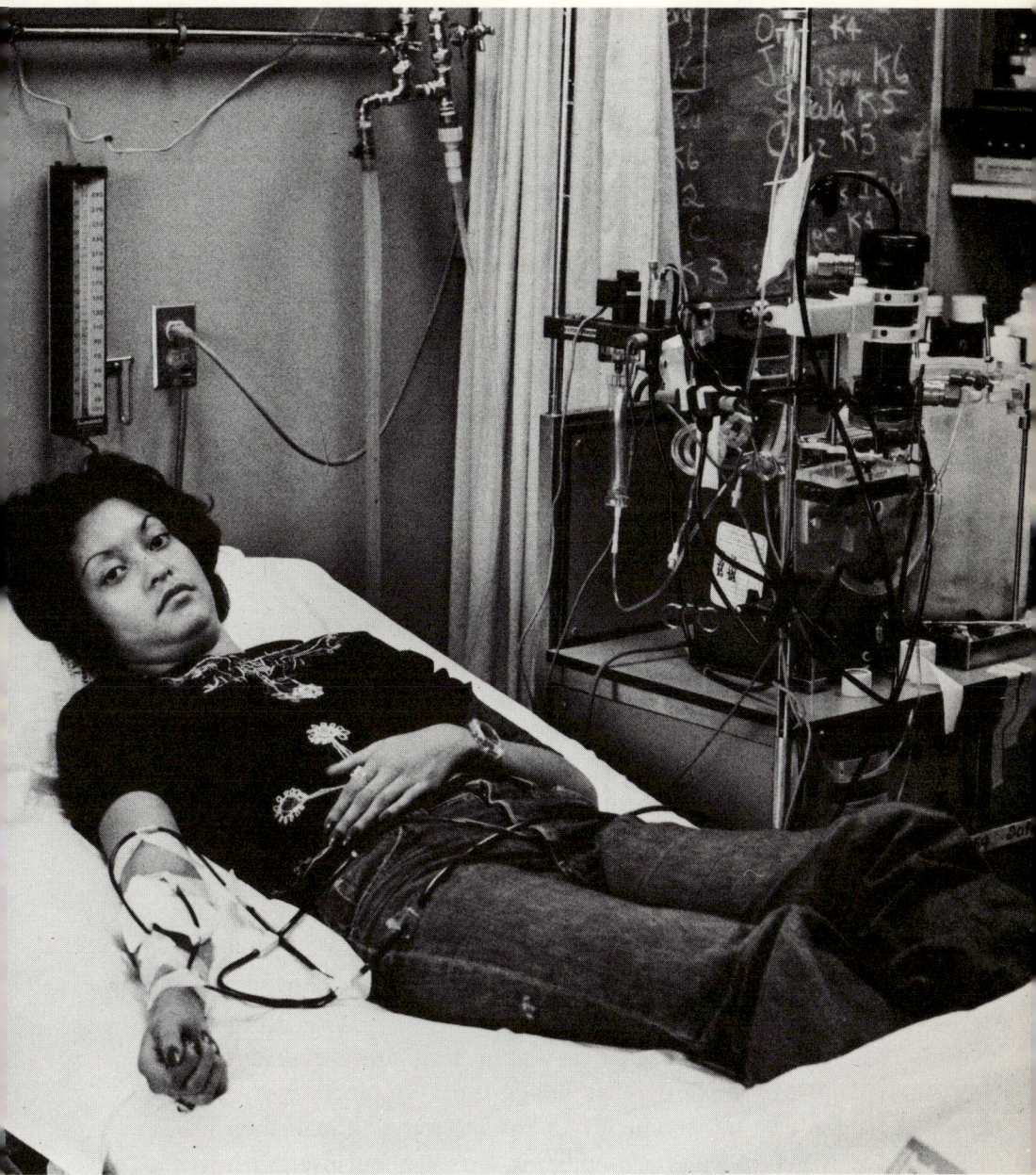

narrow tube is placed in one of the patient's arteries. The blood flows through the tube into the kidney machine. In the machine the blood passes through a long length of tubing made out of a special plastic membrane. This membrane allows the poisons and waste products to pass through, but holds back the blood cells. After passing through the machine, the blood returns through another tube to the patient's vein.

The entire blood supply passes through the machine several times. Some impurities are filtered out each trip. After a certain number of trips, depending on the patient, most of the waste products are removed.

During the treatment the patient feels no pain. It is possible to read, study, watch TV, or talk on the telephone. After the treatment the patient is able to go about his or her daily activities, looking and feeling healthy.

The artificial kidney machine has saved many lives that would have been lost because of kidney failure. But it still presents problems. The hemodialyzer, as the kidney machine is called, is very expensive to buy and operate. For patients who require treatments over many years, it is an enormous financial drain. Dependence on the machine is time-consuming and restricts travel and other freedoms. In addition, the hemodialyzer does not filter wastes from the blood nearly as well as human kidneys.

Dr. Kolff, who built the first artificial kidney, is now experimenting with a portable kidney machine. It looks like a chemistry set contained in a bulky life jacket. Someday, he hopes, it will be small enough to be truly convenient and portable.

The greater hope is for a tiny artificial kidney, implanted in the body, that will perform the function of that vital organ. The replacement may resemble the kidneys themselves, or it may be just a membrane that can be inserted into the patient's kidneys that will help in the kidneys' work.

SPARE PARTS FOR HUMANS

Arthur Gross, one-time college tennis champion, began to experience cramps in his legs whenever he walked. At the age of 42 he was forced to give up tennis because running was very painful.

An X ray examination of his legs showed that blocked blood vessels were preventing an adequate flow of blood to his leg muscles. The following year doctors operated to replace the blocked sections of arteries with artificial Dacron tubes of the same length and thickness.

By the time he left the hospital, Arthur's blood flow was normal. Within months he was back on the tennis courts free of cramps or pain. And at the age of 48, Arthur won the Minnesota State Senior Men's Singles Tennis Championship.

Every day hundreds of people all over the world receive plastic or metal tubes, plates, hinges, and joints to replace or correct damaged parts of the human body. For example:

An older woman is crippled with arthritis and as a result is unable to walk. Her doctor operates and removes the ball and socket of her hip joint. He implants a polyethylene socket into her hip bone and a stainless steel ball into the top of her femur, or thigh bone. For the first time in years she is able to walk.

A college student is in a bad car accident. A portion of his skull is smashed. The surgeon replaces the missing part of his skull with a metal plate. No permanent disability remains from the accident.

A teenager falls while skiing and badly fractures the bones in her leg. The doctor sets the broken bones into the proper position and inserts a screw through the bones to hold them together and to give added strength.

New strong, lightweight materials, plus the skills of the surgeons, have resulted in a 98 percent success rate in spare parts surgery.

ARTIFICIAL SENSES

ARTIFICIAL SENSES

ARTIFICIAL SENSES

ARTIFICIAL SENSES

THE STORY OF GEORGE REINER

George Reiner, a lawyer for a large steel company, was accidentally struck on the head by a heavy iron bar while going through the plant. He suffered a severe concussion, and was left permanently blind.

In time, George learned to read braille, to use the tape recorder, and to telephone instead of writing letters. Getting around was still difficult, however. George wanted to be independent. He did not want to use a white cane, and he did not want to depend on a guide dog.

In 1974 he purchased a laser-beam cane, especially designed for blind people. The cane sends out laser beams at three different heights to detect oncoming objects. If the head level beam strikes an obstacle, such as low-hanging branches, the cane produces high-pitched beeps; for ground level obstacles, such as a parked car, medium-pitched beeps; and for low level obstructions, such as a rock on the street or a hole in the ground, low-pitched beeps. By turning the cane as he walks, and by listening carefully to the pitch of the beeps, George is able to move around freely and independently.

ARTIFICIAL SIGHT

The laser-beam cane helps George to "see" by means of sound. It uses the same principle as bats do when they swiftly fly and swoop around, catching insects and avoiding all obstacles, even in the dark.

Bats are able to navigate entirely by sound. While they are flying they emit very short, high-pitched cries, between 10 and 20 times a second. These cries vibrate 90,000 or more times per second, which is far above the range of human hearing. With their large ears, the bats detect the echoes that bounce back from any nearby objects. The faster the return

of the sound, the closer the object. The object can be as small and as fast moving as a fly or a mosquito in flight.

In experiments at Harvard University, bats and insects were released in a large room. The bats caught the insects at an average rate of one every six seconds. When long wires were strung from floor to ceiling, and the experiment was repeated, the bats caught the insects at the same rate—and almost never bumped into the wires.

The technical name for "seeing" by sound is echolocation. In addition to bats, many sea animals use echolocation. Among them are the large dolphins and white whales, and the smaller toadfish and minnows. The clicking, creaking, and purring sounds these animals make echo back, enabling them to find their prey and avoid danger in the dark, murky depths of the sea.

Human inventors have also learned how to use echolocation to find objects under water. In sonar (sound navigation ranging), tiny pulses of sound from a sound producer aboard a ship are sent out into the water. The sound waves hit the ocean bottom, as well as any nearby objects in the water. The echoes are picked up by a microphone on the hull of the ship. An automatic calculator notes the time it takes for the sound to bounce back, and thereby measures the distance to the ocean bottom, to another ship or submarine, or to a school of fish.

Radar (radio detecting and ranging) works in a similar way. Instead of sound waves as in sonar, however, short pulses of radio waves are emitted from a radio transmitter.

**The laser-beam cane is
giving blind people
enormous mobility.**

(49)

They bounce off any surface they strike. Flight controllers bounce them off aircraft, weather forecasters bounce them off clouds, and the police bounce them off speeding cars. The reflected radio waves are picked up by a receiver that automatically changes the elapsed time into a measure of distance.

Blind people have long used echolocation to walk without help, particularly on city streets. They use the echo of their cane tapping on the pavement to learn of any obstacles in their path. Some have developed this ability to the point that they can measure distance and identify the size and shape of the obstacle by the sound of the echo and the time it takes to arrive. Blind people who have, or develop, a keen sense of hearing, and have good training, can even tell, by listening to the echoes of the cane taps, if the object they are near is made of metal, wood, or cloth.

The laser-beam cane is but one of several modern devices for the blind that use the principle of echolocation. Other devices send out audible sound waves, ultrasonic waves, or radio waves. Then, in one way or another, they pick up the reflected waves, changing the time it takes to return into an audible signal of distance to the reflecting surface.

Many of these apparatuses are large and cumbersome. A research effort is now going on to reduce their size. The goal is to make small, compact units that can be worn or easily carried. Also, researchers would like to create devices that can detect the presence of very small, as well as very large objects, and devices that can recognize the finer details of objects.

Although the eyes are the organs of sight, the actual seeing is done with the brain. In the normal eye, the light from the object enters the eye through the cornea and lens, and forms an image on the back wall of the eyeball, which is the retina. The retina is made up of many separate cells. These cells are stimulated by light and color to send tiny electrical impulses to the visual center of the brain. And it is here that the electrical impulses are changed into the images you see—the words in a book, people, trees, clouds, or whatever.

The cells of the retina, in one way, are like the dots used to create pictures in books or newspapers. If you look closely at a printed picture, you will see that it is really made up of many, many tiny dots. Where the dots are dark and close, that part of the picture looks black. Where they are smaller and farther apart, it looks gray. And where they are very small and very far apart, it looks white.

A picture forms on the retina in a similar way. The image falls on the estimated 100 million cells. Each of the cells responds to one tiny point of the total image. Some respond to whether it is dark or light, and some to color. Each cell changes this information into a tiny flow of electricity which it sends to the brain. Since there are so many cells packed so tightly together, the brain is able to blend these separate signals into a single clear, sharp image.

A bionic instrument modeled after the eye's retina is the electric scoreboard in a gym or stadium. The score is shown by lighting up a certain pattern of bulbs among hundreds of dark, or unlit bulbs. The viewer's eye and brain then blend the separate lit bulbs of this pattern into a number or letter.

This concept suggested to Dr. Willem Kolff a possible way to provide sight for the blind. His idea is to implant tiny electrodes in the visual cortex, which is the area of the brain concerned with vision.

Doctors have already carefully mapped out the visual cortex. They know exactly how each point of the visual cortex corresponds to a point in the eye's field of vision. They know that electrical charges to certain spots of the cortex will always be perceived as spots of light in the person's field of vision.

In one experiment in Dr. Kolff's laboratory, 16 wires

were attached to carefully selected sites in the visual cortex of the brain of a blind man. The wires were connected to a large computer-like piece of equipment, and to a special television camera. The computer divided the TV image into 16 segments. Each segment was either on or off. Electrical signals traveled only through the "on" wires. These signals activated the electrodes in the cortex. Each electrode created a separate flash of light that actually could be seen by the blind man. With this method, he was able to see lines drawn on a blackboard. He could even make out the rough outline of a face.

One day this method may allow the blind to read braille by sight, instead of by touch. The basic braille unit consists of six dots. Each letter is made up of some pattern of raised and flat dots. Using Dr. Kolff's equipment, the images of these dots can now be converted to flashes of light. Blind people find they are able to read braille five times faster by this visual method than by the usual touch method.

But seeing flashes of light is still not really seeing. So researchers are beginning to think about replacing a damaged eye with a sub-miniature TV camera hidden within a glass eye. Then electronic equipment, perhaps hidden inside an eyeglass frame, will convert the TV image into a pattern of on and off spots. The pattern will pass through a number of wires to the visual cortex of the brain. Here flashes of light in the brain will reproduce the image as it was picked up by the TV camera.

One of the most successful transplant operations is to replace a diseased or damaged cornea with one from a deceased human donor. Scientists hope one day to develop an artificial cornea in order to guarantee an adequate supply of this vital eye part. Some workers in the field are even exploring the possibility of transplanting an entire eye, a procedure that has not yet been done successfully.

THE STORY OF DORIS ORLET

When Doris Orlet was 16 years old she became ill with a throat infection that grew steadily worse. By the time she consulted a doctor the infection had spread from her throat to both her middle ears. She went into the hospital and, despite drug treatments, ran a very high temperature for almost two weeks. When she finally recovered she had suffered a permanent loss of hearing in both ears.

The sudden and complete loss of hearing seriously depressed the girl. She tried various treatments to improve her condition. None offered her much assistance. Therefore, when Doris learned of an experimental device that could, in some cases, return hearing to the deaf, she was very interested. The doctors in charge of the research agreed that Doris would be a good subject for their studies.

In a delicate operation, they attached electrodes to Doris' inner ear. Wires from the electrodes were connected to a miniature microphone that was implanted in the side of her skull. When they hooked up the apparatus there was no need to ask Doris if she could hear. Her shriek of joy was proof enough that they had created an artificial ear that worked.

ARTIFICIAL HEARING AND SPEECH

Hearing is the second most important of the senses. Because it is very sensitive, it is easily damaged by infection or injury. Serious damage may result in deafness.

The normal ear works somewhat like the normal eye. Sounds enter the outer ear and pass through to the inner ear. Sensing cells in the inner ear convert the vibrations of the sounds into nerve impulses. These go along nerve paths to the auditory center in the brain where sound is perceived.

In the experimental artificial ear, the sounds are picked up by a miniature microphone. The microphone may be implanted in the surface of the skull, just behind the outer ear, or hidden in eyeglass frames. The microphone changes the sounds into electricity, just as in the normal ear. The electrical current is fed through wires to the inner ear, which passes it on to the brain. For people with certain kinds of damage in the outer or middle ear, this promises some degree of hearing.

Although speech is not one of the senses, it is closely related to hearing, and scientists are working to develop an artificial voice, just as they are improving artificial hearing.

In the past, people who had their vocal cords removed because of some disease could only speak by what is known as "burp" speech. That is, they swallowed air, and then formed the words as they burped up the air.

Dr. Stanley Taub of New York Medical College is developing a voice improver to give those without vocal cords a better means of speaking. He has invented a device that sends air directly into the windpipe. It consists of a small case that is worn on the upper chest. From the case a tube leads to an opening in the neck and then into the windpipe. The pressure of the air sets the tissue of the esophagus into vibration, producing speech that sounds almost normal.

The first, small steps towards artificial sight, artificial hearing, and artificial speech have already been taken. Hopefully we can come even closer to producing more faithful imitations of the human models in the near future.

ARTIFICIAL INTELLIGENCE

ARTIFICIAL INTELLIGENCE

ARTIFICIAL INTELLIGENCE

ARTIFICIAL INTELLIGENCE

ARTIFICIAL INTELLIGENCE
AT WORK

You dial a telephone number that has been changed and you hear a voice say: "The number you have called, 4.3.2.2.3.6.4., has been changed. The new number is 4.3.2.7.3.9.8."

You approach a supermarket entrance, and find that the doors open up even before you push them open.

On an air flight, the pilot and co-pilot leave the cockpit and head toward the rear of the plane. The plane is flown by an automatic pilot until they return.

There are trains that go along their regular route, making all the stops and opening and closing the doors, without anyone on the trains driving or controlling them.

At home or in school on wintry days, you notice that the heat starts coming up as soon as the temperature of the air grows a bit chilly. And then, when it warms up, you find that the heat stops, and the air begins to cool off.

In a restaurant juke box, you see a mechanical arm pick out a record, place it on a spinning turntable as a tone arm comes down and plays the records.

There are cameras that automatically adjust so that they always take pictures with the correct exposure, whether in a dimly lit room or out in the bright sunshine.

Some cars now produce a buzzing sound if you sit down without attaching the seat belt.

Most soda machines dispense a can of soda and also the proper change.

These are just some examples of machines that behave in "intelligent" ways. Others launch rockets and make airline reservations, run machines in factories and keep track of bank accounts, measure the heartbeat and blood pressure of patients in hospitals, play chess, and other games, and much more.

ARTIFICIAL INTELLIGENCE MACHINES

To the scientists in this field of artificial intelligence, intelligence is defined as having a memory, an ability to do calculations and make decisions, and the equipment to work reasonably fast.

Human intelligence, of course, goes far beyond artificial intelligence. It includes verbal, numerical, and spatial reasoning. It includes memory and the ability to use remembered past experiences to help in understanding new experiences. It is the ability to grasp an entire complex situation from just a few small cues. And it is the curiosity to seek answers and explanations for things that are not easily understood.

Although artificial intelligence is much more limited than human intelligence, machines have been created to do a number of extremely complex tasks. In fact, there are a number of tasks that are done by machines better, faster, and with fewer errors than by humans.

Basically, all artificial intelligence machines are divided into two parts. One part receives information from outside the machine. This corresponds to the human senses of seeing, hearing, feeling, and so on. The other part processes this information, making any calculations, comparisons, and decisions that are necessary. This corresponds to the human brain.

The part of the artificial intelligence machine that receives outside information is often based on a sensor. Sensors change one form of energy, such as heat, light, sound, or pressure, into another form, usually either electricity or movement.

A phonograph is built around a sensor in the tone arm. As the needle goes through the record grooves, it moves back and forth, up and down, according to the sounds that have been recorded. In many phonographs, the needle is attached to a special crystal in the tone arm. As the needle moves, it pushes and pulls on the crystal, creating tiny electrical charges on the surface of the crystal. In this way, the phonograph changes the movement of the needle into electrical signals.

The speaker of the phonograph is also a sensor—a sensor that changes electricity into sound. It takes the electrical energy from the crystal, builds it up, and passes it to a metal coil placed in a magnet, causing the coil to vibrate. A paper cone attached to the coil also vibrates, and the vibrations of the paper cone produce the sound.

There are sensors that change light into electricity. They are called photoelectric cells. When light strikes the surface of the cell, it causes electricity to flow; when the light ceases, so does the flow of electricity.

And there are sensors that change heat into movement. The bimetallic thermometer consists of strips of two different metals, usually brass and steel, joined together along their length. Both metals expand when they are heated, but the brass expands more than the steel at the same temperature. This causes the strip to curl, with the expanded brass on the outside of the curve. The higher the temperature, the greater the curling movement of the metal strip.

In some artificial intelligence machines the sensors are used in the simplest way. The automatic door opener, for example, is based on a photoelectric cell. As you approach the door, you pass between two posts. In one post, a source of light is aimed to shine on a photoelectric cell in the other post. Normally the light strikes the cell, and electricity flows. That keeps the door shut. When someone walks between the posts, however, that person blocks the light. The flow of electricity ceases, and the door automatically opens. After the person passes through, the light strikes the cell again, and the door is automatically closed.

(59)

In the seat of a car that automatically sounds a buzzer when you do not attach the seat belt, is a sensor that changes pressure into electricity. When you sit down, you press on a strain gauge, producing an electrical current which sounds the buzzer. When you pull on the seat belt to put it on, you break the electrical circuit, and the buzzer stops.

In more complicated machines the sensors are part of a feedback system. In a feedback system the information received by the sensor is compared with certain facts or numbers that are stored in the machine. The machine acts after it compares the outside information with the stored facts and numbers.

The thermostat that controls the heat in a room is an example of a machine that works on the feedback principle. The heart of the thermostat is a bimetallic thermometer. When you set the thermostat at a comfortable temperature, say 68°F. (20°C.), you are automatically adjusting the metal strip. When the temperature is 68° or above, the strip is curled. This breaks the electrical circuit that controls the oil burner and the oil burner goes off. But as soon as the temperature drops below 68°, the strip straightens out, thus completing the circuit and restarting the oil burner. The oil burner stays on until the temperature rises to 68°. Then the process repeats itself.

In the most complicated artificial intelligence machines the sensors send their information to computer-like processors. These machines are built around immense numbers of electrical switches and other electronic devices. They are bionic imitations of the nerve cells of the human brain.

The computer is one of the most amazing inventions of our time. It can solve almost any problem, as long as the problem can be broken down into small steps, and can be stated in exact terms. It can solve mathematical equations and perform complicated calculations. It can answer questions in logic, provided the questions and concepts can be stated in exact symbolic form. It can compare different sets of numbers or values. It can store vast amounts of information, and instantly retrieve this information from its memory. In addition, it performs all these functions at fantastic speeds, far faster than any human.

Playing chess by computer puts to use the computer's immense memory, its incredible speed, and its ability to make decisions in highly complex situations. In order to get a computer to play chess, each square on the board and each chess piece is given an identifying number. Then a numerical value is assigned to each square and each position.

The game is actually played by using a computer terminal that looks like an electric typewriter hooked up to the computer. The human player types out his or her moves, naming the piece and telling the square to which it is moved. The computer makes its moves by automatically typing out the corresponding information on its moves.

A good chess player, human or machine, will think through three or more moves in advance. To do this correctly, the computer must be programmed to examine over 1,000,000,000 possible combinations of positions before making a move!

A large, well-programmed computer, with a big memory capacity, can outplay all but the top-ranked chess players. In time the chess champion of the world may no longer be a man or a woman, but a computer.

The telephone company uses the artificial intelligence of a computer to tell you that a number has been changed. Usually when you dial a number, you trip a series of switches that automatically send your call through to the correct number. But if the number has been changed, the number you dial goes instead to the memory of a computer.

The computer memory contains all the numbers that

have been recently changed. In a flash, the computer locates the old number, and with it, the new number. With equal speed, the computer searches through a collection of pre-recorded tapes of all the digits and various words. It selects and plays the tapes that, when played one after another, tell you the number you have reached, and the new, changed number.

The best computers have good memories, outstanding ability to solve problems without error, and work at great speeds that far surpass their human models. But, as mentioned earlier, these characteristics are just a small part of the much more complex human intelligence. Artificial intelligence is still limited to repetitive functions, to problem-solving in a step-by-step progression, and it can only do what humans have built it to do.

**Scottish chess master
David Levy challenges the
CYBER-176 supercomputer
to a game of chess.**

(63)

ROBOTS
ROBOTS
ROBOTS
ROBOTS
ROBOTS

THE STORY OF
ROBBIE THE ROBOT

The household robot stood in the corner of Mr. and Mrs. Kane's kitchen. A plug from its base was connected to the electrical outlet in the wall.

"Clear the table, Robbie," said Mrs. Kane, as the family finished dinner. There was a clicking sound and a flashing green light went on as the robot started gliding toward the table on its padded wheels. As it moved, a small television camera on top of the human-sized robot slowly turned back and forth.

When it reached the table, Robbie stood still while the camera scanned the surface. Then its two arms extended and picked up the plates one by one, gently stacking them in a tray.

"Load the dishwasher," commanded Mrs. Kane, who was now sitting on the living room sofa. At the sound of these words, the robot clicked again, flashed its light, and slowly turned, bringing the tray of dishes to the dishwasher. While the family read and listened to music on the phonograph, Robbie loaded the dishwasher. On further instructions from Mrs. Kane, it swept the floor. Then, when there was no more work to be done, the robot silently glided back to the corner. Its plug worked its way back into the electrical outlet to be recharged. The clicks stopped, and the light went off.

ROBOTS:
STATE OF THE ART

Today, of course, this story of Robbie the Robot is science fiction. But tomorrow, or a few years from now, it may be an accurate description of how the work gets done in a typical kitchen.

Scientists in laboratories in many universities, including those at the Massachusetts Institute of Technology and Stanford University, are doing research into understanding the theory of robots and into solving the practical problems of building robots. And, according to a 1973 survey, there are some 100 factories around the world that are actually manufacturing robots.

The standard definition of a robot is a machine that performs tasks as a person would. Or, put another way, it is a mechanical person. The word itself was not coined until 1923, when it was first used by the Czechoslovakian playwright, Karel Capek. It comes from the Czech word, *robot,* which means work or drudge.

The makers of today's robots had to get rid of some old ideas. Robots are not Frankenstein monsters, bent on attacking and destroying anyone or anything nearby. Instead, most robot-like artificial intelligence machines are helpful, and work for the benefit of people.

Also, robots need not look like humans with human-appearing head and body, arms and legs. The term, androids, describes certain robots purposely made to resemble humans.

Robots probably will be approximately the same size as humans, since they must fit into, and be able to work in, an environment that is scaled to human dimensions. But they will not necessarily look like humans.

To build a robot, you need a thorough understanding of all the theories behind the construction of a sensing, think-

**The robots from STAR WARS,
C3PO and R2D2, are not as
fanciful as they may seem.
Robots are a fact of the present.**

(68)

ing, and doing artificial creature. This includes a grasp of bionics and cybernetics: How is information from the environment received and processed? How are comparisons, decisions, and choice of appropriate actions made? And, how are actions carried out?

Suppose, as an example, you wanted to build a robot that could play a game like blindman's buff. First, you would have to answer certain theoretical questions: How would you teach the game to someone who had never played before? How would you explain the object of the game? What suggestions would you offer to the person who is blindfolded to help him or her find one of the other players? Exactly what steps would you tell the person to follow in order to find his or her way around the room? Then, once these questions were answered, you would have to assemble and combine all the hardware to carry out the necessary processes and actions.

Professor Meredith Thring, of Queen Mary College, England, has prepared a minimum list of features that every robot should have. The list includes: an arm and hand, a self-powered propulsion system for mobility, a computer for memory and decision-making ability, and sensors that can recognize and measure pressure, temperature, sound, light, weight, shape, size, color, distance, as well as receptors for radio waves and electrical impulses.

The mechanical arm and hand for a robot can be built along the same lines as an artificial limb. But building a robot's arm should be even easier, since the scientist need only be concerned that it performs well, and not that it resembles a human arm.

Human fingers, for example, are able to do a variety of tasks because the tips are soft and flexible, allowing them to grip and hold extremely well. One type of industrial robot, called the Versatran, can be fitted with different hands depending on the task it will be doing. One set has inflatable plastic pads on the fingers. With this hand, the Versatran can securely grasp objects of irregular shape.

Almost all industrial robots in use so far perform their tasks while fixed in one place. Those that are mobile run along a fixed path. They are guided either by tracks or by an electrical conductor that is buried in the floor. One problem with a fully mobile robot in a factory is that a fault in the steering or guiding system might cause the robot to run into some working piece of machinery, damaging or destroying both the machine and the robot.

In some instances an electric motor is used to power the wheels of a fully mobile robot. The motor runs on batteries, and the batteries are recharged by plugging the robot into an electrical source. For situations where movement by wheels is not satisfactory, 2-, 3-, or 4-legged walking robots are being developed. Such robots would have the advantage of being able to go up and down steps and over rough terrain.

At the heart of most advanced robots is a computer. And while computers exist that can direct mobile robots, they are usually too large to be placed within a robot, and too expensive for practical use. The latest computers, however, are much smaller and cheaper than at anytime in their history. For example, there are now computer memory devices that are only 20-millionths of an inch (50-millionths of a cm.) in diameter. There are computer electrical circuits that are so small that they can only be installed with the help of a microscope.

Most of the sensors needed by a robot already exist. Strain gauges measure pressure, photoelectric cells measure light, microphones measure sound, thermometers measure heat, and so on.

With what we know of the theory of robots, and with the development of the different parts that go into making robots, how far have we gone in actually building robots?

(71)

ROBOT TURTLES

The turtles named Elmer (*electro-mechanical robot*) and Elsie (electro *light sensitive with internal* and external stability) are outstanding figures in robot history. They were built during the 1950s in England by Grey Walter, along with a second-generation robot named Cora (conditioned *reflex* analog).

More recently robot turtles have been built at the Aston Cybernetics Laboratory in England, and at the Artificial Intelligence Laboratory of the Massachusetts Institute of Technology.

Scientists have found that building a robot turtle is the easiest and least expensive way to test their knowledge and skill with robots. The typical robot turtle is about 12 inches (30 cm.) long from front to back, with a large, hunched shell that extends up some 8 inches (20 cm.). It has three wheels, one in front and two at the back. The front wheel has two motors—one to drive the wheel, the other for turning and steering. The shell is movable and has several built-in pressure sensors. On the front of the shell are two photoelectric cells; inside is a microphone.

Robot turtles are designed to head toward a weak source of light, and away from a powerful source of light. The photoelectric cells direct its movements. If its batteries are running down, though, the robot will head toward the powerful light, since the electric source to recharge its batteries is located there.

The movements of the robot turtles may seem strange, but they do follow a logical plan. Suppose a turtle is moving toward a light, and it strikes an obstacle. The robot will stop, back away, and start forward again, slightly off to the side. If it encounters the obstacle again, it stops, backs away, and starts forward once more, further along. Although it looks hit-or-miss, it is really part of a careful plan. The turtle will make a full circle, if necessary, to get past the obstacle.

(72)

If someone blows a whistle while the robot is in motion, it stops in its tracks. But if the whistle is blown several times at the same moment as the robot picks up a weak light, it will react to the whistle as though it were a light, and will proceed straight ahead.

But what are sophisticated scientists trying to learn by teaching a robot to turn toward or away from a light?

Their work is actually enhancing our ability to build general robots that will be useful to people. A robot that heads toward a source of light may, one day, be developed into a robot fire fighter that moves toward a fire and directs a stream of water at it. Such a robot would lessen the endangerment of human fire fighters.

Also, in setting these goals for the robot researchers are finding out more about the human mind and human learning. They are becoming good observers of people. How do people find their way around obstacles? How do they find the shortest route to a target goal? To build a fine robot, the scientists must first find answers to basic questions about humans.

MASTER-SLAVE MANIPULATORS

A laboratory in Texas is doing research on a deadly type of virus. If this virus should get out into the air, it would kill anyone who breathed it in. In the laboratory it is kept in a sealed, airtight room. Yet the scientists must be able to place it in test tubes with various kinds of animal cells and with combinations of chemicals, and then observe the results.

A large hospital in Michigan uses radioactive chemicals in very small amounts to treat some diseases. These are extremely dangerous chemicals. If doctors or laboratory workers were exposed to these chemicals, they would suffer a fatal dose of radiation. Yet these chemicals have to be mixed and prepared for use on the patients.

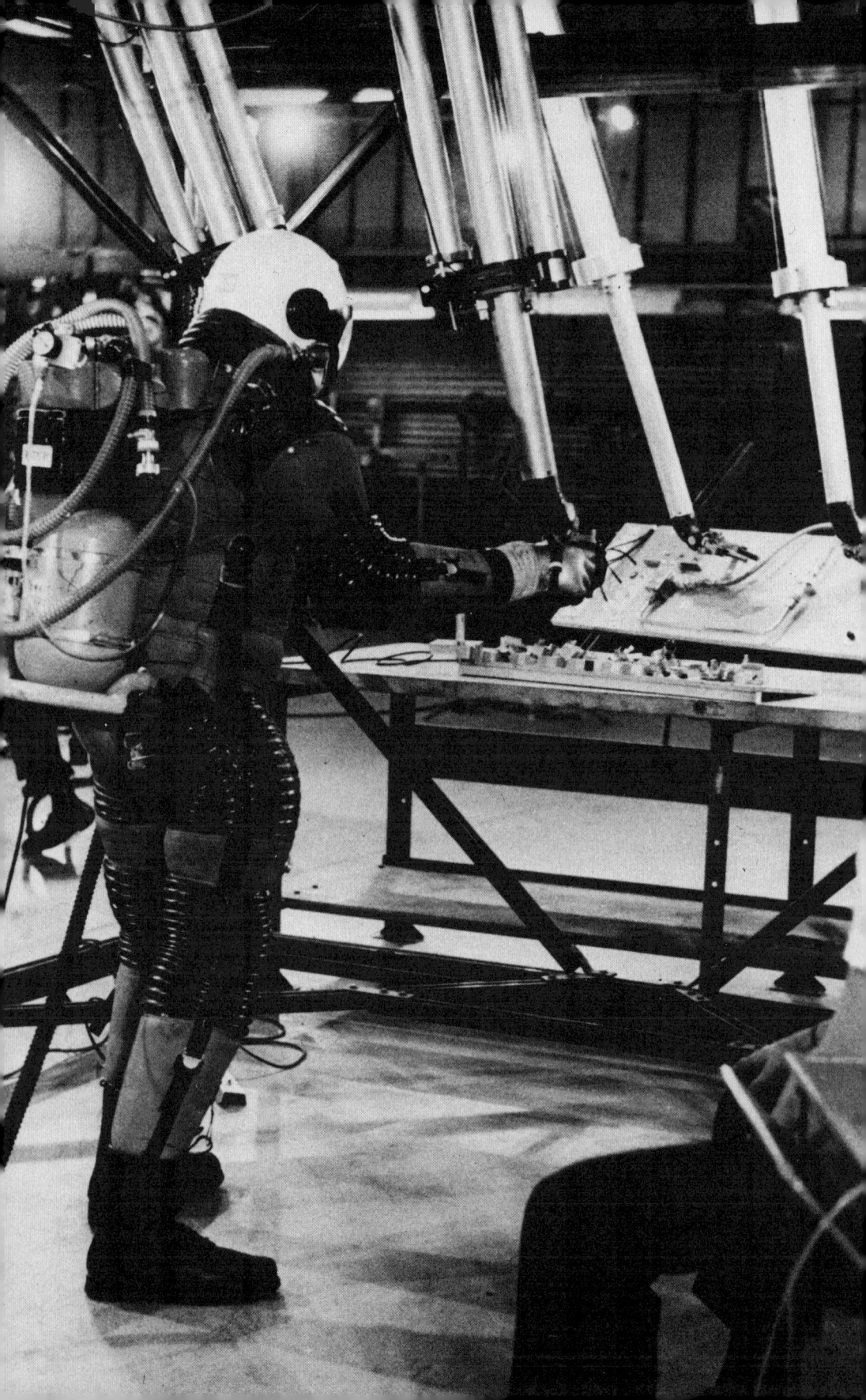

In both of these places, and in many other laboratories, hospitals, and factories all around the world, robot-like machines—called master-slave manipulators—allow workers to do important research with dangerous materials. The workers control these manipulators from behind walls that protect them from the harmful substances. Only the devices make contact with them. The workers see what they are doing either by looking through a large glass window set in the wall, or by watching on a closed circuit television screen.

In its simplest form there is a direct link from the human controls to the robot's arms and fingers in the master-slave manipulator. When the worker raises an arm, the manipulator's arm goes up. When the worker clasps the controls with his or her fingers, the metal fingers of the manipulator come together.

The more advanced master-slave manipulators have motorized arms and fingers. The motors make the movements larger and more powerful than the movements of the workers who operate them.

Perhaps the most advanced master-slave machine is one that was developed for the U.S. Army some years ago. It is called CAM, (cybernetic anthropomorphic *machine*). It is designed to increase both the strength and range of a person's arms and legs.

The CAM has two arms and two legs, each one about 6½ feet (2 m.) long. An operator sits in the cab above its legs. As the operator's arms and legs move, the CAM mirrors the motions with its arms and legs. However, the CAM movements are four times as large and four times as powerful as the operator's movements.

**A master-slave manipulator
is being tested for its
feasability in space use.**

The CAM has walked over very rough terrain, and has stepped over 4-foot (1.2-m.) obstacles. It has sidled through narrow pathways, sidestepping objects in its way. It has lifted the front of a jeep with one leg, and pushed a jeep along by placing one foot inside the car. And it has carried a 500-pound (226.5-kg.) load in its arms.

WORKING ROBOTS

Most robots in use today either do repetitive tasks in a factory, or are used for space exploration.

The simplest industrial robot is nothing more than a mechanical arm. Typically, it brings raw materials to a machine to be processed, and then removes the processed items. The hand of the mechanical arm might work by gripping the material between two metal fingers, by a magnet to pick up metal objects, by a hook if there is a convenient hole, or by suction if it is sheets of glass, metal, or plastic.

The more advanced industrial robots have three parts— an arm and hand, an eye, and a brain. In most cases the arm and hand is a sort of artificial limb, the eye is a television camera, and the brain is a computer.

An experimental Japanese industrial robot, called the HIVIP Mark I, has two TV camera eyes. One camera focuses on three pictures showing different angles of a structure made up of blocks of different sizes. The other camera scans over the surface of a table on which the blocks have been scattered.

The images from the first TV camera go to the computer, which uses the images to calculate how the blocks are arranged to form the structure. The images from the second camera also go to the computer, so that the computer can learn the location on the table of each of the different blocks.

The computer is then able to direct the HIVIP's arm to select the blocks in the correct order and stack them so as to duplicate the pictures seen by the first camera.

(76)

A similar robot at Stanford University has been programmed to construct a water pump from separate parts. It gathers all the parts together, assembles them, and screws them into place.

On July 20, 1976, the robot that was part of the Viking I space mission landed on the planet Mars. Because of the great distance (some 50 million miles or 80 million km.), because of the low temperatures (down to –122°F. or –85°C.), and because no one was sure what dangers might be present, it was decided to send a robot instead of a human astronaut.

Radio signals from Earth directed the robot to go through the various operations it was programmed to do. It photographed the Martian landscape. It measured temperature, wind speed, humidity, atmospheric pressure, and other weather conditions. It was equipped to sense any earthquake activity, even though this part of the robot did not work. And it used its mechanical arm to scoop up samples of dirt and deposit them in various containers inside the robot. Moreover, it put these samples through various chemical and biological tests to learn the chemical composition of the soil, and to seek signs of life. Finally, it radioed back to Earth the results of all its activities.

The robots that exist today are quite remarkable in what they can do. Even more remarkable is the promise of what they will be able to do in the future, as yesterday's bionic fiction becomes today's bionic facts.

INDEX

Androids, 68

Armes, Jay, 15, 17

Arms, artificial, 17–22

Artificial bones, 3, 9

Artificial cornea, 52

Artificial hearing, 9, 12, 53–54

Artificial heart, 37

Artificial intelligence, 5, 12, 57–61, 63

Artificial kidney, 40, 42

Artificial limbs, 3, 5, 9, 12
    arms and hands, 17–22
    cost of, 17, 20
    feeling and, 20–21
    legs, 22, 24–25
    myoelectricity, 19–20
    nuclear energy and, 3, 21

Artificial sight, 47, 49–52

Artificial ventricle, 36–37

Aston Cybernetics Laboratory, England, 72

Bats, 9, 47, 49

Bimetallic thermometer, 59, 60

Biologists, 5

Biomedical engineering, 11–12

Bionics
    artificial limbs, *see* Artificial limbs
    brain and, 38–40
    cybernetics, 10–11, 70
    defined, 5

(79)

Heart disease, 29, 35
Heart-lung machine, 35, 36
Heart valve, 36
Hegesistratus, 6
Hemodialyzer, 40, 42
Hilton, Reid, 19
HIVIP Mark I, 75

Icarus, 6
Industrial robot, 70–71, 76
Intelligence, artificial, 5, 12, 57–61, 63

Kantrowitz, Adrian, 36
Kennedy, Teddy, Jr., 22, 41
Kidneys, 40, 42
Kolff, Willem, 37, 40, 42, 51, 52

Laser-beam cane, 47, 50
Legs, artificial, 22, 24–25
Lens, 50
Leonardo da Vinci, 6

Massachusetts Institute of Technology, 68
Artificial Intelligence Laboratory, 72
Master-slave manipulators, 73, 75–76
Mathematicians, 5
Mechanical pumps, 36
Microphones, 71
Myoelectricity, 19–20

Norse legends, 6
Nuclear energy, 5, 11
artificial heart and, 37
artificial limbs and, 3, 21
pacemakers and, 29, 31

Open-heart surgery, 35–36
Organs, 9, 12
heart, see Heart
kidneys, 40, 42
Orlet, Doris, 53
Oxygen, 31

Pacemaker
brain, 38–39
heart, 29, 31, 35
Painkiller device, 39–40
Persia, 6
Photoelectric cells, 59, 71, 72–73
Physicians, 5
Plutonium rays, 32
Prostheses, defined, 17
Prosthetics, defined, 17

Radar (radio detecting and ranging), 9, 49–50
Radiation, 21, 32, 73
Reiner, George, 47
Retina, 50–51
Revere, Paul, 9
Robot turtles, 72–73
Robots, 67–68, 70–73
computers and, 71, 76

ABOUT THE AUTHOR ABOUT THE AUTHOR ABOUT THE AUTHOR ABOUT THE AUTHOR

Melvin Berger, author of over
three dozen books for young readers,
is a school teacher, musician,
editor, and antiques collector.
He lives in New York State,
with his two daughters and
author-wife Gilda Berger.